We'll run and jump on the grass.

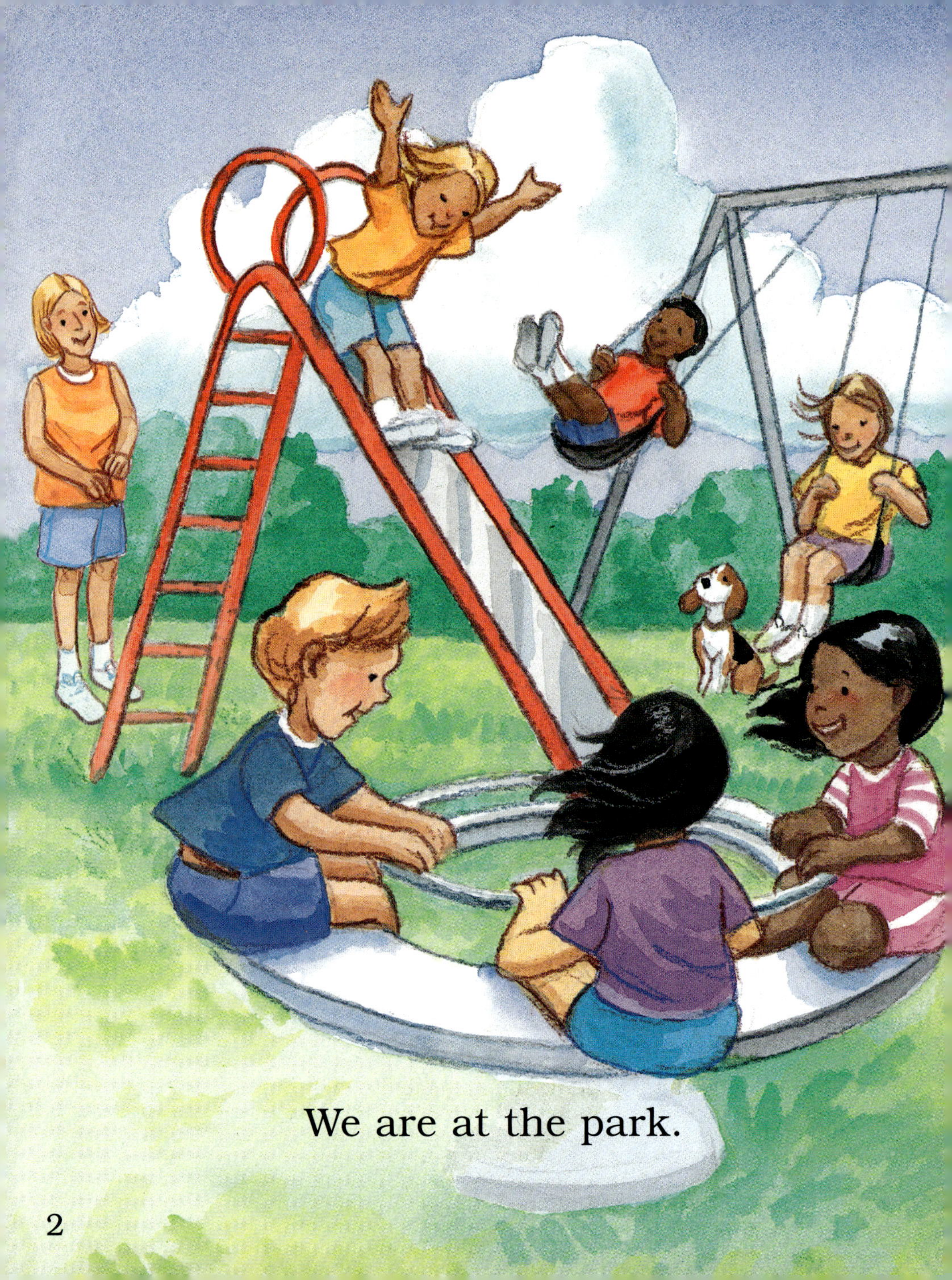

We are at the park.

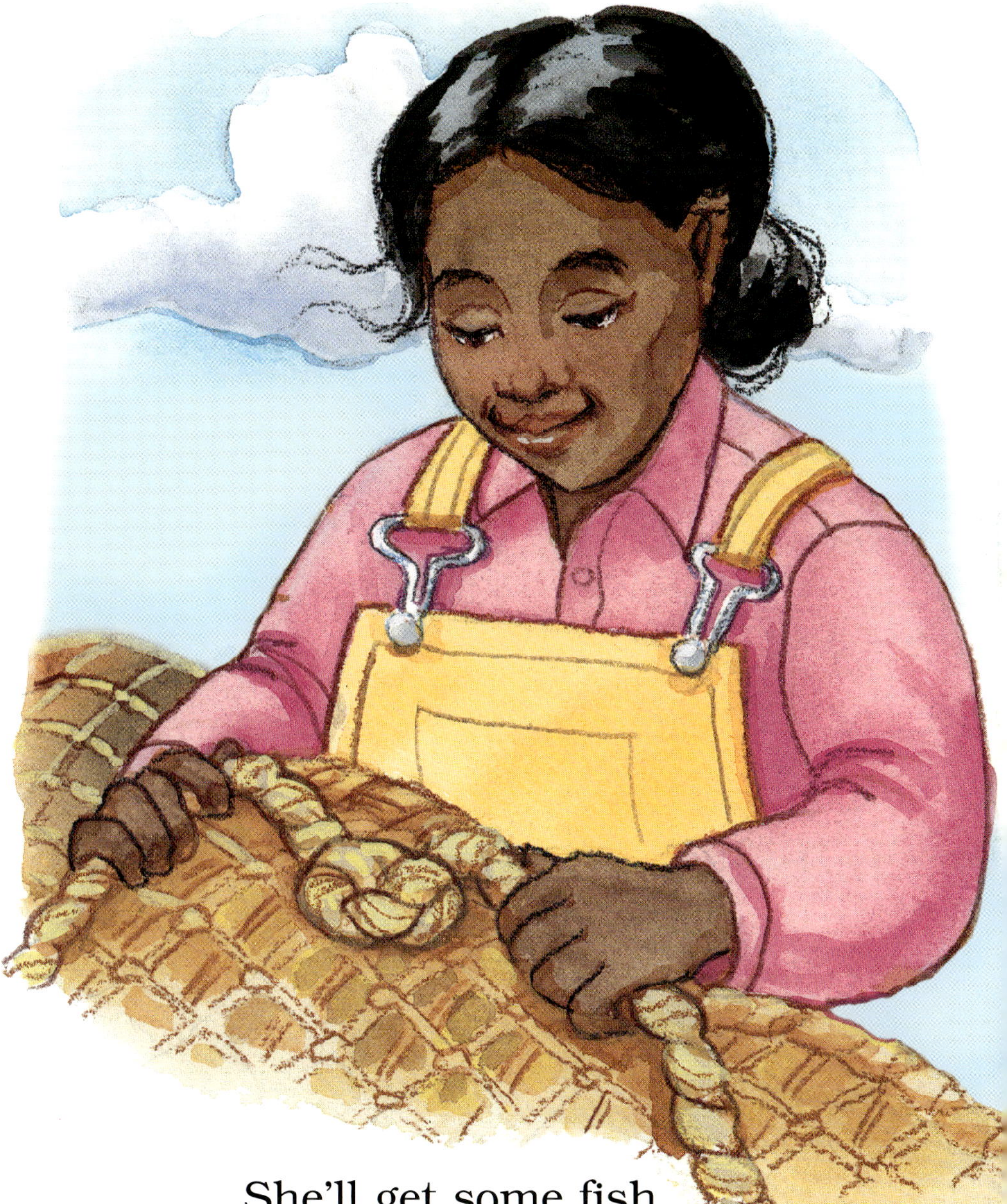

She'll get some fish.

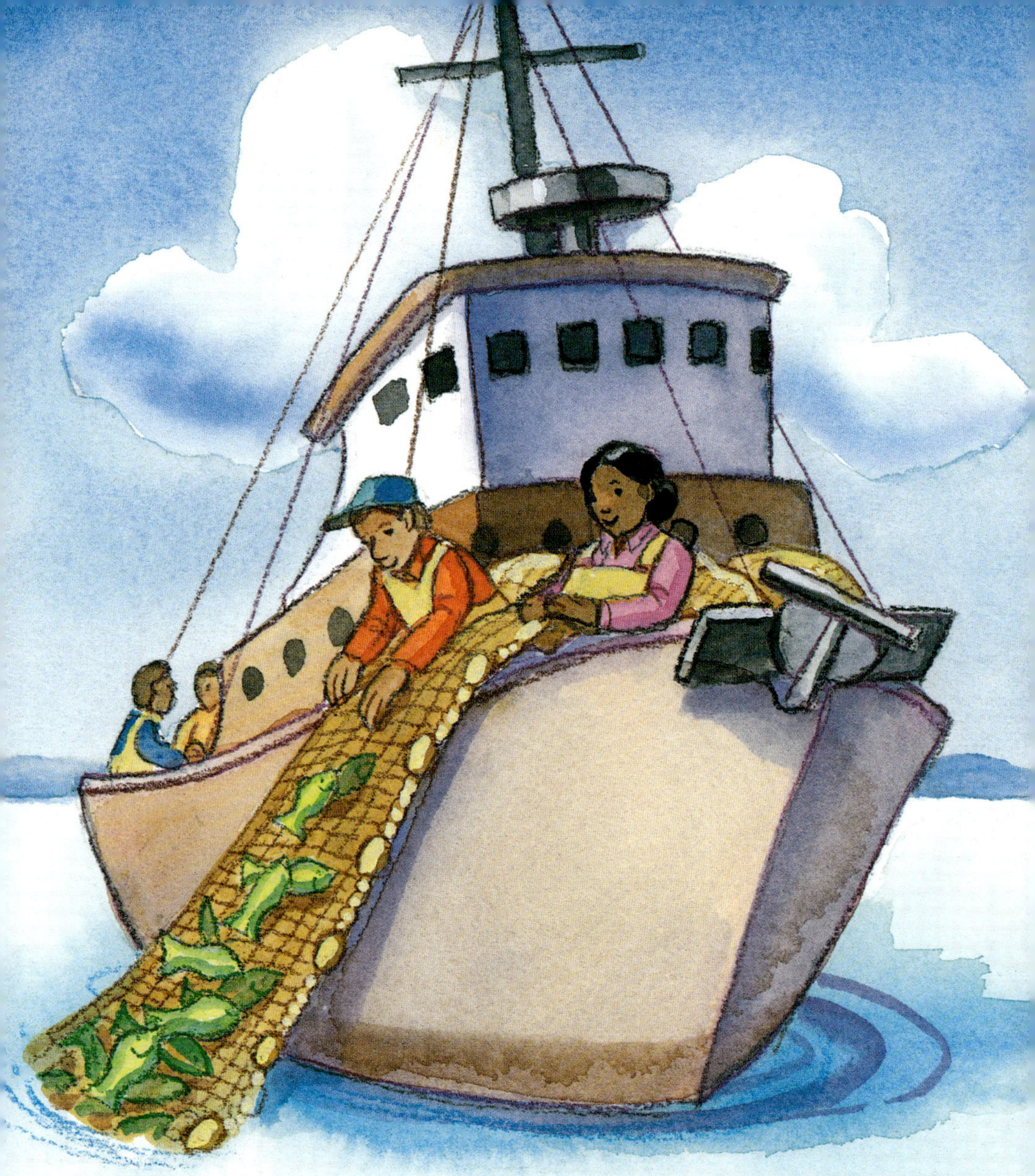

She's hard at work on a ship.

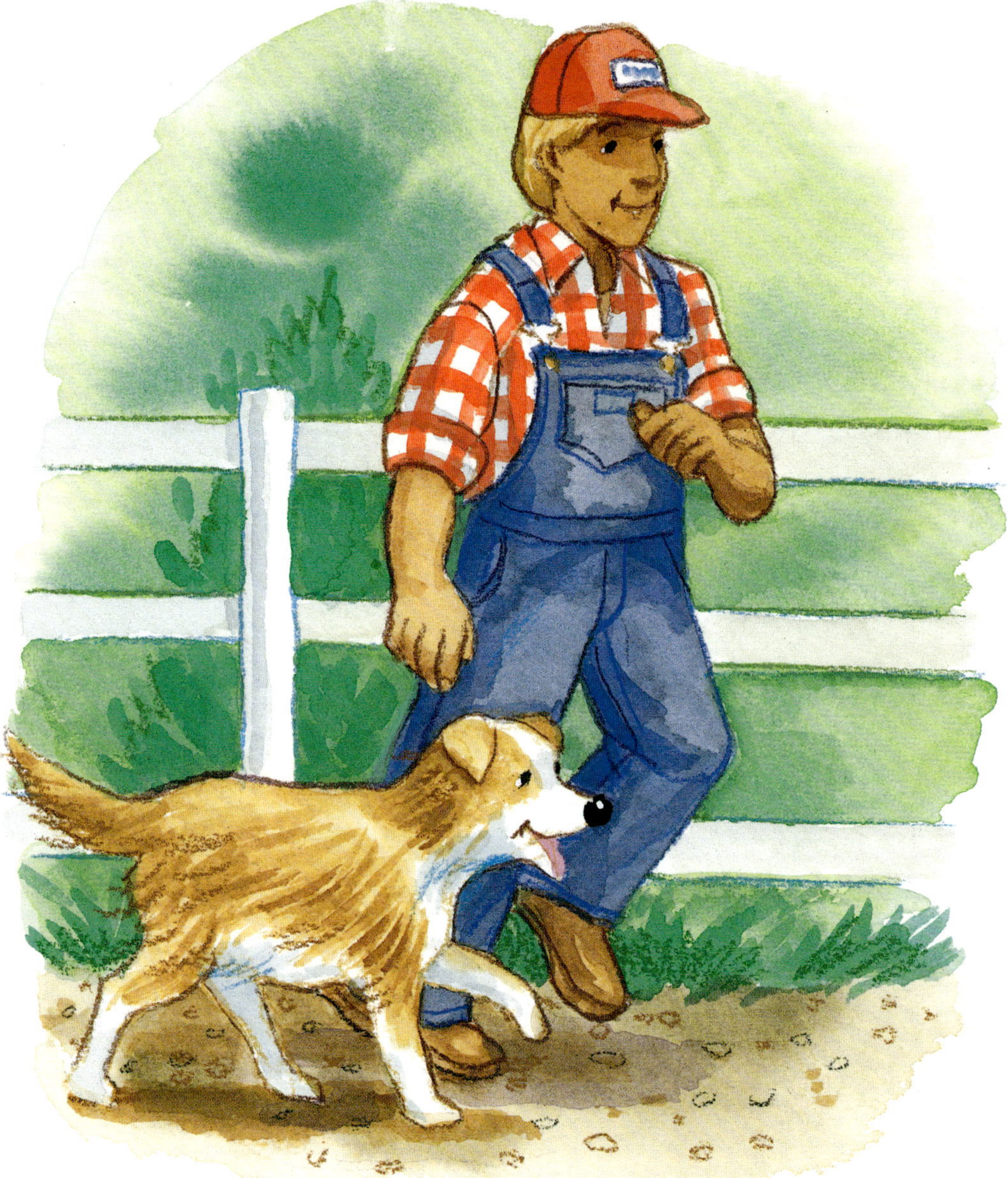

He'll go to the barn.

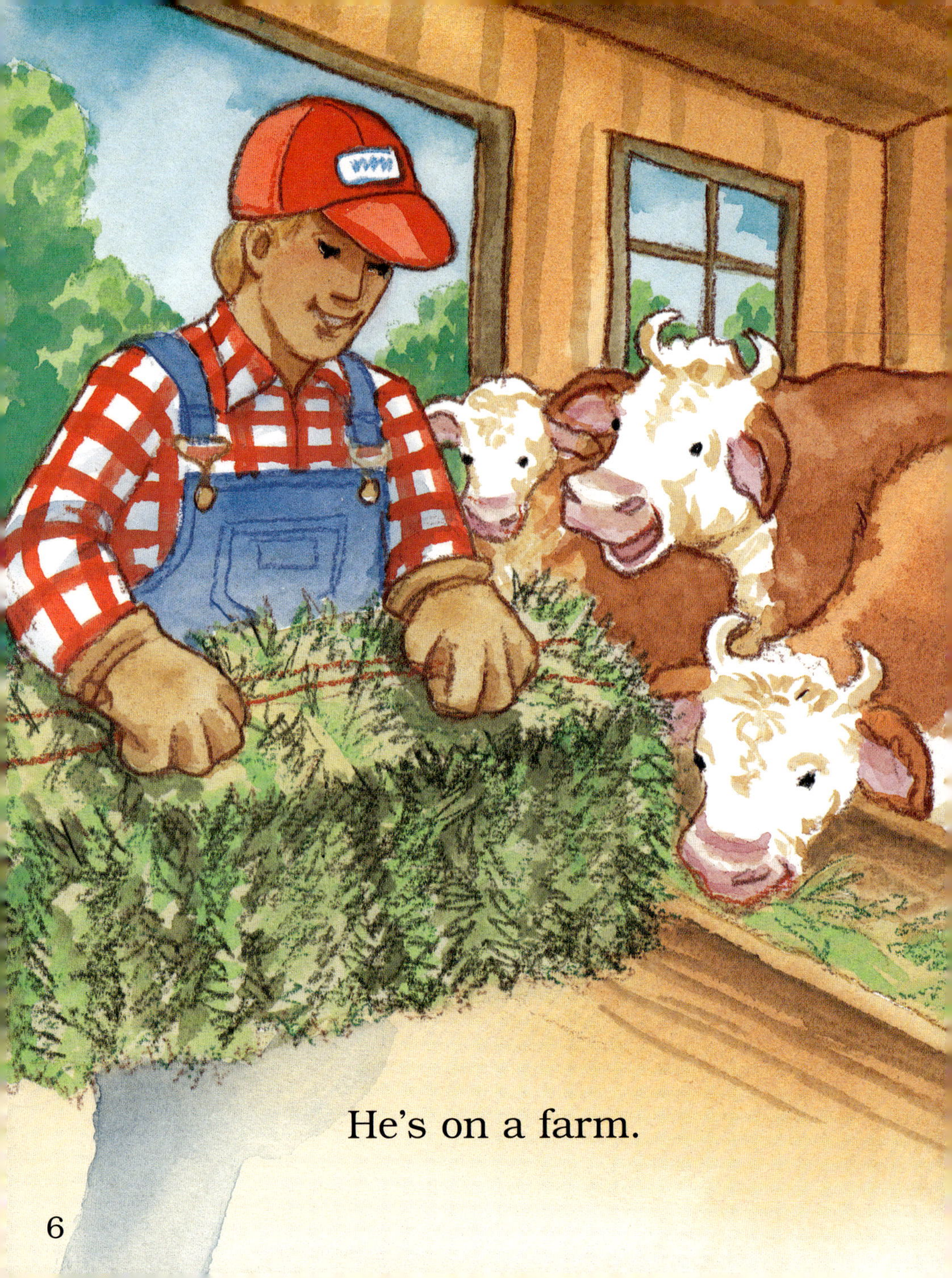

He’s on a farm.

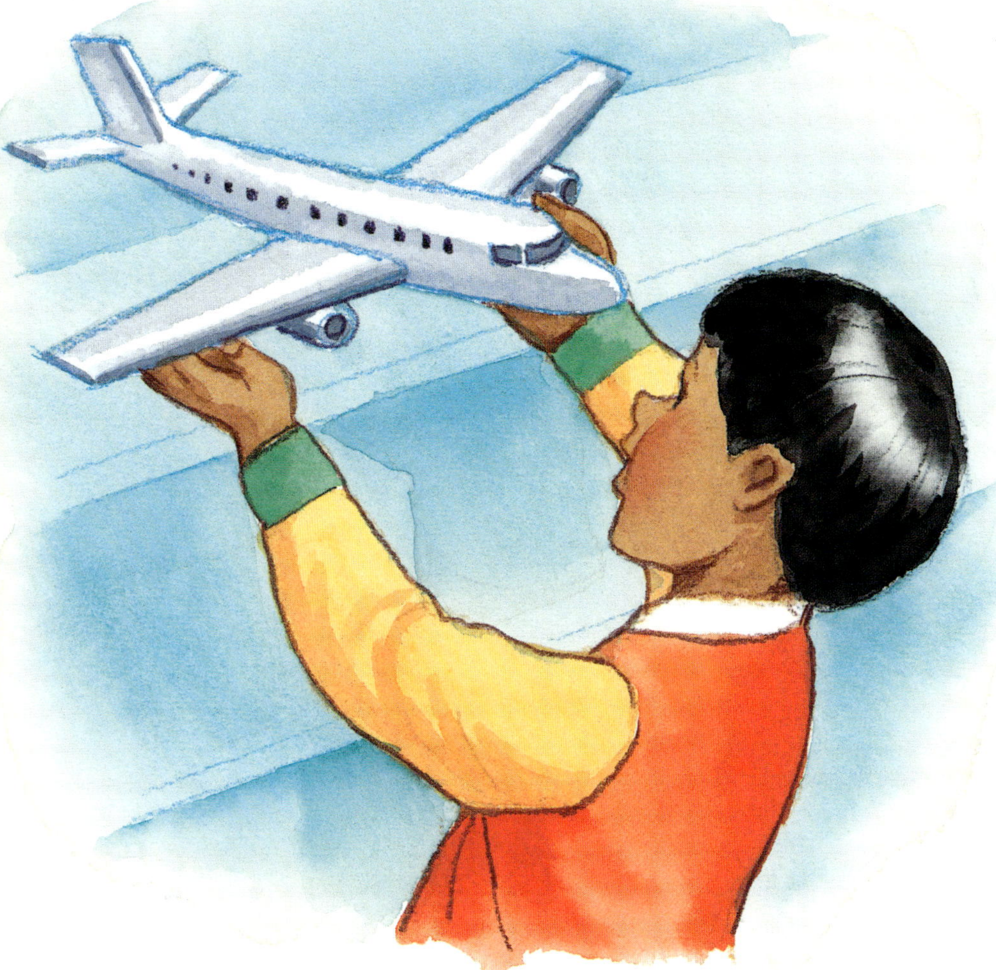

I'll get the jet.

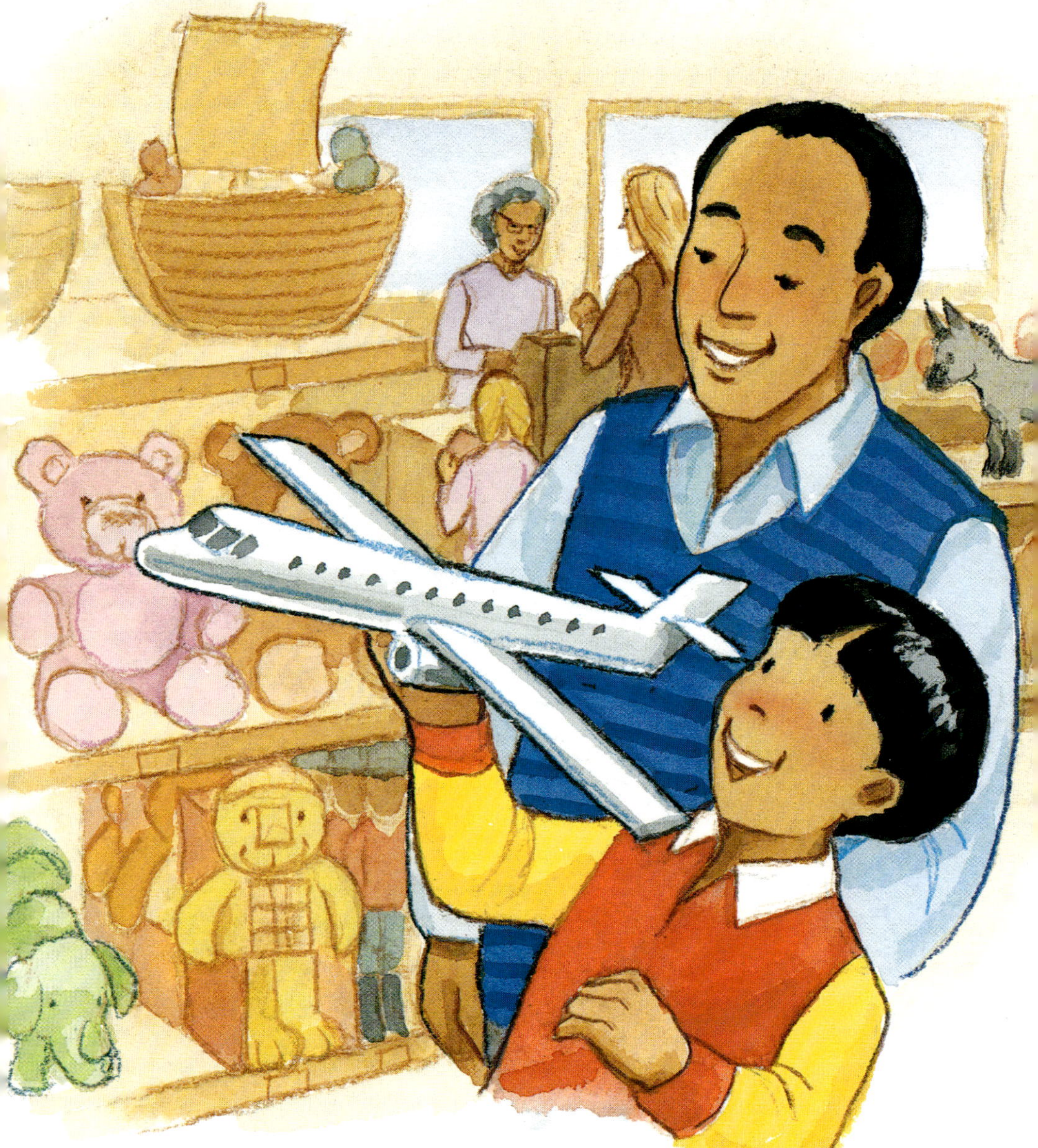

Where am I?